This book was made for the purpose of my Senior Seminar class, but also for fun. I would like to thank my high school Senior Seminar main advisor Karen Morlan, my On-Site advisor for my book Kristina Jenny, Yia Yia (grandma) for giving me ideas for the book, my sister Aubri for letting me borrow her iPad for the drawings, and my mother and friends for supporting me.

A story about a girl who doesn't care for bugs or the environment

Written and illustrated by Acacia Rhodes and her anonymous friend

Phoenix doesn't care because of the BUG in her Hair!

Phoenix was asleep.

Then her mom came in before her alarm clock went

BEEP!

She told her that she had a big day ahead

So she woke her up for some French toast made with bread.

She felt something on her forehead tickle.

It felt kind of funny,

Because there was a bug as big as a

nickel.

As soon as she saw it, she screamed in despair.

"Oh no! There's a bug in my hair!"

She ran out of the room as fast as she could.

But the bugs followed her out past the door.

Phoenix hates bugs, she thinks they are no good.

But her mom is worried now because she caused an uproar!

After getting dressed,
She ran to the kitchen to get away.

She seemed very stressed,
She told her mom what happened and said, "I hate bugs in every way!"

Her mom identified it as a Cicada, which comes once every 17 years.

But Phoenix said, "I don't care about bugs or the earth!" and ran outside in tears.

A little boy named Dash, the neighbor next door,

Saw Phoenix was upset,

So he came over to help her not fret.

The Cicadas were loud,

Almost as loud as a tiger that roars.

When he came outside, she said hi.
He asked if she wanted to play, and look for bugs with him all day.

But she said, "No! No way!"

He also asked if she wanted to go with his family to the zoo.

She responded, "I'll go, but only because I want to hear the cows go moo."

Phoenix, Dash, and his family hop in the car.

They ask him, "Are we there yet?"

His dad says, "It isn't that far."

Phoenix responds, "Okay, bet."

First, they decide to go to the store.
To get some food, for a picnic outdoors.

Dash's father chooses plant-based food,

and when Phoenix saw it she said "Ewe!"
So he explained how it's better for the animals and the planet,

but she says, "I'd rather it just vanish!"

They arrive at the park to sit and eat,

Phoenix says how she wishes she had meat.

Dash's father says, "What a sight, with all the bugs and lovely wildlife."

But Phoenix doesn't care for the earth because of the bug that was in her hair!

They finally get to the zoo.
Well-fed and ready to see all the animals, they yell,
"Woo Hoo!"

They can hear the Cicadas here,

they are louder than at home.

The noise tells them that they're near.

Phoenix says, "Cicadas are as loud as trombones!"

As they walk over to the education classroom,
The screen begins to play a movie

One about climate change.

Dash says, "That's pretty groovy!"

They learn about the planet and how it's getting worse.

It's caused by us humans, and cannot be solved with a nurse.

A visit to the Polar Bears was one to remember.

If they are not saved,

they won't be here each December.

They are put in the zoo for conservation,

so if they are preserved, we won't have to say "Salutations!"

When they got home,
it was starting to get dark.

The sun was setting,
it looked like a work of art.

They all had fun and will get plenty of sleep,
And hopefully, dream of a flock of-

Sheep.

Phoenix's mom says, "goodnight, little one,

Your big day is finally done."

So she went into her bed,

when suddenly, the bug flew back in over her head!

This made Phoenix both sad, and mad.

So she said, "I hate bugs, they're all bad; they don't matter anyway and never have!"

And then she wished that bugs never existed,

then, when she fell asleep,
everything shifted...

So she dreamed a good dream

Or so, she thought.

That bugs never existed,
Which was kind of mean.

As she fell asleep she dreamed of a world where butterflies, bees, spider, and cicadas

don't creep.

In her dream, she went back to the store.
However, it wasn't the same anymore.

All the fresh fruits and vegetables were gone.

There was nothing to eat, not even an apple or beyond.

Without our bugs, we would have no fruit.

Or barely anything to eat, except, maybe dairy and meat.

If we had no bugs or people that cared for the earth,

There would be no chance for a rebirth.

All of our parks would be polluted, and our trees would die.

Eventually, we would all have to say

goodbye.

Not just our land would be affected,
But also the wildlife.

Animals wouldn't be protected.
Because of this, lots of species would die.

Just like the Polar Bears would go extinct.
And this is why.

Oh No!

When Phoenix woke up,

she realized her mistake.

It was a nightmare, not a dream.

And it was NOT a piece of cake.

Phoenix runs to her mother in tears.

She realizes her mistake,

that she has to get over her fears.

She tells her "I'm sorry" and that she was wrong.
Bugs are good for the earth,

And they **belong**.

About the Author

Acacia Rhodes previously known as Acacia Longnecker was born in Tucson Arizona in 2004. She grew up in Missoula Montana and attended Hellgate Elementary until she moved in 2014. The red house within this book replicates her house that she grew up in Montana (shown on left). After her parents got divorced she moved 2000 miles with her mom and her two sisters (on the left) to Columbus Ohio in order to be closer to her grandparents (shown on right). She is a huge environmentalist who ran as Co-president for environmental club at her school. She is currently a senior at her local high school and is attending a university in the fall to pursue her dream of studying veterinary and environmental science, and continuing to educate others about climate change.

www.ingramcontent.com/pod-product-compliance
Lightning Source LLC
Chambersburg PA
CBHW051833210526
45473CB00005B/1861